So You Want to Learn About

Birds

Written and Photographs by
Katrina Willoughby

I hear you want to learn about birds!

You probably see these winged creatures almost every day. Where ever you live, you can find birds near you. They live all across the world.

When you mention birds, most people think about feathers.

Birds are the only living animals with feathers. The number of feathers a bird has varies by species. Song birds can have as few as 1,500 feathers while some eagles have as many as 8,000 feathers.

This Black Swan is native to Australia. Swans can have up to 25,000 feathers.

Over time a bird will molt. Old feathers fall out and new feathers grow in their place.

Have you seen a duck preening itself? Ducks and other birds spend a lot of time straightening their feathers and making them waterproof.

Near the tail, birds have a gland that makes oil. They use their beak to move and work this oil into their feathers to become waterproof.

Most birds fly, but there are a few birds like penguins and ostriches that don't fly.

For all the birds that do fly, feathers are shaped and positioned to be aerodynamic. Some feathers can be intentionally moved to help with flight. To fly birds change their wing position similar to the way we move our hands and arms when swimming.

You may have seen a bird moving its tail feathers to change direction or slow down.

The way a bird's body is shaped makes it aerodynamic and helps it fly.

A bird's wings are wider at the front and taper toward the back. Wings have a longer curve across the top. This means that air has to travel a longer distance over the top of the wing than the bottom. Air under the wing is pushed together to make it denser and push the bird up.

Flying takes a lot of energy. Using up so much energy means birds must eat a lot.

A bird's beak is designed for how it eats. This wood-pecker has a long, strong beak to peck into wood and take out insects.

A hook on the end of the beak of this Crested Caracara helps it hunt small animals.

A bird is warm blooded just like humans. However, to have enough energy to fly, they have to have a higher body temperature than us. The average body temperature of birds is 105 F (40 C). This is even higher than a dog's temperature.

To keep warm in the winter a bird will build up fat in the fall so they have extra insulation. Just like us, a bird will shiver when it gets cold. They will fluff up their feathers and tuck in their bills to keep warm.

Some birds sun themselves by stretching out their wings to feel the heat of sun rays. Below an Anhinga spreads its wings to warm.

Even a bird's feet and legs are designed to keep it warm.

Specialized scales on the bird's feet will help prevent heat from leaving them. They can constrict blood flow to their feet to keep the rest of their body warmer. When it is hot out some birds like vultures will defecate on their feet to cool off.

Do you know how many toes a bird has? The majority of birds have 4 toes.

Many birds that perch on branches have 3 toes in the front and 1 in the back. This lets them wrap their toes around branches and have a tight grip.

Parrots also have 4 toes, but they have 2 pointing forward and 2 pointing backwards.

Many hunting birds have talons.

The long toes of this Wattled Jacana allow it to walk across lily pads floating on the water.

Birds that swim use their feet more like paddles and may have webbed feet. A duck will have 3 webbed toes in front and one toe in back.

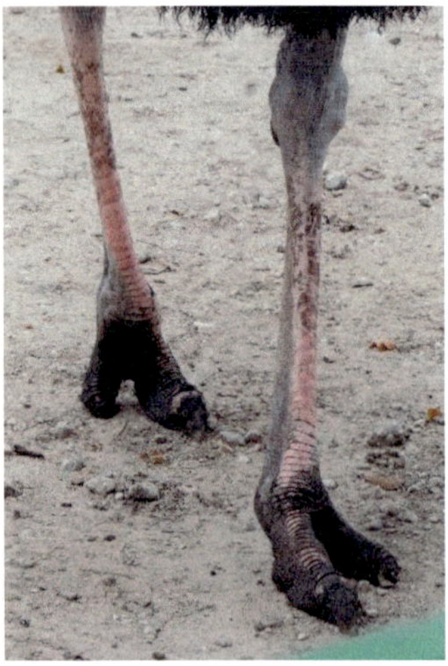

Ostriches are the heaviest bird, and they don't fly. Only two strong toes are found on each foot.

Ostriches can live up to 40 years, be 9 feet tall, and weigh over 300 pounds. Strong legs are used to kick predators like lions and let this large bird run up to 45 miles per hour.

In cartoons you may have seen an ostrich bury its head in the sand. They don't really do this. They do lay low and press their neck and head to the ground to try to hide from predators.

Ostriches have larger eyes than any other land animal. Its eye is bigger than its brain!

Birds can see a lot better than you and me.

Eagles can see small animals up to 2 miles away!

Birds like owls are awake at night. This is called nocturnal. Their night time vision is excellent.

While most birds have their eyes looking to the sides, owls have their eyes looking forward. To change where they are looking, they have to move their head. Some owls can turn their heads to look behind them.

Not all birds make nests. Those that do are making a safe place and lay and raise their young.

Many nests are made of branches and leaves, but those aren't the only things used to make a nest.

The shape of the nest and what it is made of will depend on the species of the bird.

The duck above used her own feathers and wood mulch to make her nest.

There are even birds that make their nests out of their saliva! They are called Edible-Nest Swiftlets and are found only in Asia. Their nests are considered a delicacy and people actually eat the nests!

For many species, it is the male's responsibility to build the nest. For others, the male and female will build the nest together. This Black-Bill Magpie and its mate can take up to 40 days to build a large nest of sticks that can be 3 feet tall.

Here you can see a swallow's nest. It is made of mud and sticks.

Below you can see the open mouth of a baby Swallow reaching up to get food from its parent.

That baby hatched from an egg.

Birds do not lay all of their eggs at once. Instead, they may lay one a day or one every few days until the clutch is laid.

The chickens commonly used to lay eggs you buy at the grocery store can lay one egg a day for up to two years!

After laying the eggs the parents will incubate them to keep the eggs at the right temperature for the babies to grow. This often means sitting on the eggs to keep them warm.

A group of the same species of birds is called a flock. Some birds live in flocks all year long to provide better protection against predators. Others may flock together only during part of the year to migrate or to spend the winter together.

Not all birds flock together. Some species will primarily live in pairs. Others may spend much time alone.

Robins can be very territorial and live a solitary life for much of the year.

Have you ever been pooped on by a bird? It looks different than poop made by mammals. This is because birds don't pee and poop separately. Their solid and liquid waste is combined before it leavcs their bodies.

Birds come in all different colors. Generally, the males are more colorful than the females.

The top picture shows a female Cardinal while the bright bird at the bottom is a male Cardinal.

Flamingos get their bright pink color from their food. The bright pink shrimp and algae they eat have pigment (color) that gets passed to the flamingo.

So....

Did you learn new things about birds?

I hope you learned new things and enjoyed this book!

This is the sixth book in the series "So You Want to Learn About..." Please check out previous books including the first book "So You Want to Learn About Butterflies" and check back for new books.

Thank you for reading.

Other books in this series

So You Want to Learn About Butterflies

So You Want to Learn About Turtles & Tortoises

So You Want to Learn About Reptiles & Amphibians

So You Want to Learn About Insects & Bugs

So You want to Learn About Spiders

Made in the USA
Las Vegas, NV
07 November 2026

57669927R00017